Yetis

by Grace Hansen

Abdo
WORLD OF MYTHICAL BEINGS
Kids

Abdo Kids Jumbo is an Imprint of Abdo Kids
abdobooks.com

abdobooks.com

Published by Abdo Kids, a division of ABDO, P.O. Box 398166, Minneapolis, Minnesota 55439.

Printed in the United States of America, North Mankato, Minnesota.

102023

012024

THIS BOOK CONTAINS RECYCLED MATERIALS

Photo Credits: Alamy, Getty Images, Granger Collection, Shutterstock, ©Fandom p22/CC BY-SA 3.0

Production Contributors: Teddy Borth, Jennie Forsberg, Grace Hansen
Design Contributors: Candice Keimig, Pakou Moua

Library of Congress Control Number: 2023937688

Publisher's Cataloging-in-Publication Data

Names: Hansen, Grace, author.

Title: Yetis / by Grace Hansen

Description: Minneapolis, Minnesota : Abdo Kids, 2024 | Series: World of mythical beings | Includes online resources and index.

Identifiers: ISBN 9781098268619 (lib. bdg.) | ISBN 9781098269319 (ebook) | ISBN 9781098269661 (Read-to-Me ebook)

Subjects: LCSH: Yeti--Juvenile literature. | Abominable snowman--Juvenile literature. | Mythical animals--Juvenile literature. | Folklore--Juvenile literature. | Legends--Juvenile literature.

Classification: DDC 398.2454--dc23

Table of Contents

Myth of the Yeti 4

Mega Mountain Man 8

Protectors 18

Yetis Today 20

More Cryptids. 22

Glossary 23

Index 24

Abdo Kids Code. 24

Myth of the Yeti

Yetis are creatures from **folklore**. Stories of similar creatures, called cryptids, are told in many **cultures** around the world. Some of these tales go back thousands of years.

After years of stories, the **Sherpa** people began calling the creature "*Yeh-Teh*." Europeans called it "Yeti." In the 1920s, the British named the creature the "abominable snowman."

Mega Mountain Man

Yeti **lore** began in the Himalayas. The **Sherpa** people live in Nepal and Tibet. They have stories about a large, hairy beast.

Himalayas
N
W
E
S

Yeti is described as an ape-like creature that walks on two feet. It is usually said to have reddish-brown fur. Some people described it as having yellow, grayish-white, or black hair.

Reports put Yeti anywhere from 5 to 10 feet (1.5-3 m) tall. People also say Yeti has a skunk-like odor.

Yetis are believed to be **carnivores**. They eat animals **native** to the mountains, such as marmots. They may eat larger animals, such as yaks.

Yetis are **nocturnal** creatures. This is why they are most often seen at night. They are said to make grunting and growling noises.

Protectors

The **Sherpa** people believe Yetis to be guardians of the forest. The creatures must be treated with respect.

Yetis Today

Tales of Yeti and other cryptids remain popular today. Yetis appear in books, television, and movies. Films, including *Smallfoot* and *Abominable*, feature Yetis.

More Cryptids

Glossary

carnivore – an animal that only eats meat.

culture – the language, customs, ideas, and art of a particular group of people.

folklore – the stories and ways of a group of people from a certain place or country.

lore – something that is learned from traditional knowledge or belief.

native – being the place of birth or origin.

nocturnal – active at night.

Sherpa – a member of the Tibetan people living in Nepal who are known for their expertise in mountain climbing.

Index

Abominable (movie) 20

appearance 8, 10, 12

behavior 16

British people 6

color 10

cryptids 4, 20

Europeans 6

food 14

Himalayas 8

media 20

Nepal 8, 18

noises 16

Sherpa people 6, 8, 18

size 12

Smallfoot (movie) 20

smell 12

Tibet 8, 18